Growing with Mathematics

Student Book

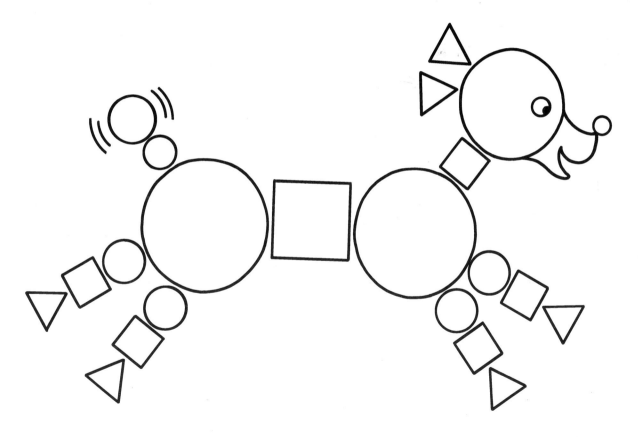

Contents

Getting Started
Color your favorites 3
Draw your favorite center activity 4

Topic 1: Sorting and Classifying
Sort by one attribute 5
Match leaf shapes 6

Topic 2: Counting to Ten
Count and draw 7–12

Topic 3: Space, 3-D Shapes, & Patterns
Use positional language 13–14
Recognize 3-D shapes 15–16
Create patterns 17–18

Topic 4: Exploring Numbers 0 to 9
Match quantities and numerals 19–20
Match quantities and number words 21
Match quantities and numerals 22
Write numerals 23–25
Match quantities, number words, and numerals 26

Review Week A
Count and color quantities 27
Match quantities and numerals 28
Match 3-D shapes 29
Extend and create patterns 30
Match quantities and number words 31
Write numerals 32

Topic 5: Introducing Measurement
Show comparative size 33–34
Identify change 35–36

Topic 6: Comparing, Ordering, & Joining Numbers
Identify more and fewer 37–38
Write numerals 39–40
Match quantities and numerals 41
Write numerals to match quantities 42–43
Analyze the numbers six to nine 44–47
Combine two groups 48

Topic 7: Exploring Patterns and 2-D Shapes
Work with circles 49
Work with triangles 50
Identify 3- and 4-sided shapes 51
Work with squares 52
Extend and create patterns 53
Identify squares, circles, and triangles 54

Topic 8: Grouping and Separating Numbers
Focus on ten 55–59
Write missing numbers 60
Match quantities, number words, and numerals 61
Combine two groups 62
Identify penny, nickel, and dime 63
Combine two groups 64

Review Week B
Show comparative size 65
Extend and create patterns 66
Match quantities, number words, and numerals 67–68
Combine two groups 69–70

Topic 9: Exploring Numbers 11 to 15
Match quantities and number words 71–72
Match quantities and numerals 73–76

Topic 10: Measurement
Measure length 77–78
Record weather data 79
Order events 80

Topic 11: Exploring Numbers 0 to 20
Match quantities and numerals 81
Match quantities and number words 82–83
Match quantities and numerals 84–85
Write missing numbers 86
Combine two groups 87–88

Topic 12: Equal Groups, Sharing, & Fractions
Work with equal groups 89–90
Share to make equal groups 91
Identify halves 92

Review Week C
Identify left and right 93
Count and record data 94
Make equal groups 95
Review numbers 1 to 20 96

Name ....

Color your favorite crayon. | Color your favorite ice cream.

Ask, **What is your favorite color? Make the crayon that color**.
Ask, **What is your favorite ice-cream flavor? Color the ice cream to match**.

Getting Started   **3**

Draw your favorite activity.

Name _____

Revisit *Dinosaurs at School*. Then talk about the centers in your classroom. Ask children to draw something from their favorite center.

Color the things you **wear**.

Talk about the pictures on the page. Say, *Which are the things you can wear? Put a counter on them and then color them orange.*
Find the things you cannot wear. Color them blue.
Extension: Ask, *How else could you sort the things on this page?* (For example, into groups of things you can eat and things you can play with.)

Find the pairs of leaves that match.
Make them the **same** color.

Ask, *Can you find the pairs of leaves that match? Make them the same color.*
Extension: Give children a sheet of paper. Have them draw two things that are
alike and two things that are **different** from each other.

Draw creature faces.
Count the 👁 and ☺.

Name Jan

Display the *Ten Funny Creatures* song poster, then tell the children that they are going to make their own funny creatures.
Encourage children to draw several eyes or mouths on each creature.
After they have completed their creatures, ask, **How many creatures are there? How many eyes did you draw? How many mouths?**

Draw a  for each penguin. Count the 🪁.

Name _____

Display the penguin illustration from *The Animal Parade*. Then say, **Count the penguins on your page. How many are there? Now give each penguin a kite. How many kites did you draw?**
Extension: Have children draw another kite for each penguin and then count all the kites again.

Draw a ⬭ for each duck. Count the ⬭.

Name ....

Display the duck illustration from *The Animal Parade*. Then say, ***How many ducks are there on your page?***
Count them. Now give each duck a balloon. How many balloons did you draw?
Extension: Have children draw another balloon for each duck and then count all the balloons again.

Draw a on each 👆.

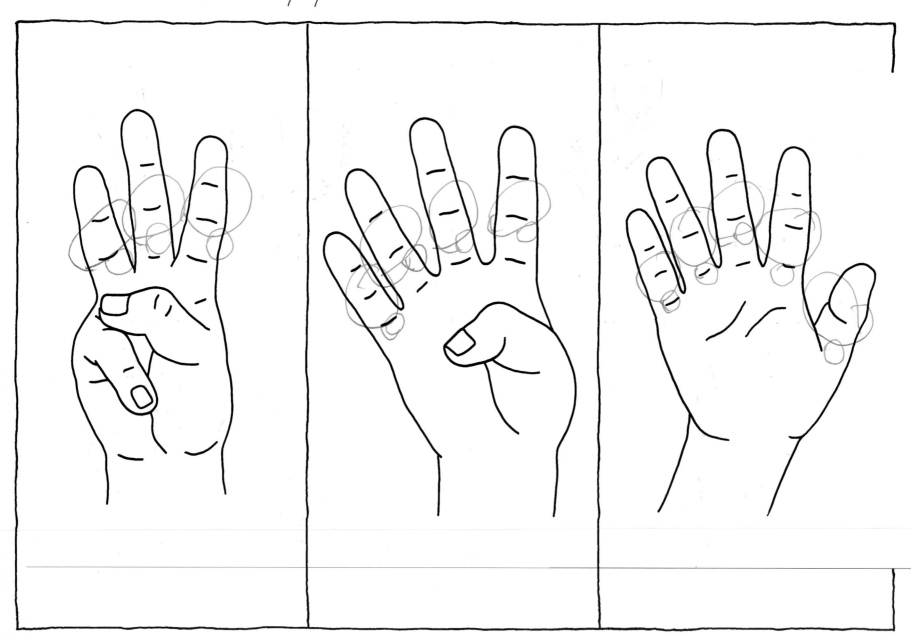

Say, **Look at the first hand. Count the fingers that are standing up.**
Now draw a ring on each finger that is standing up.
Repeat for the other two hands.

Draw a on each 👆.

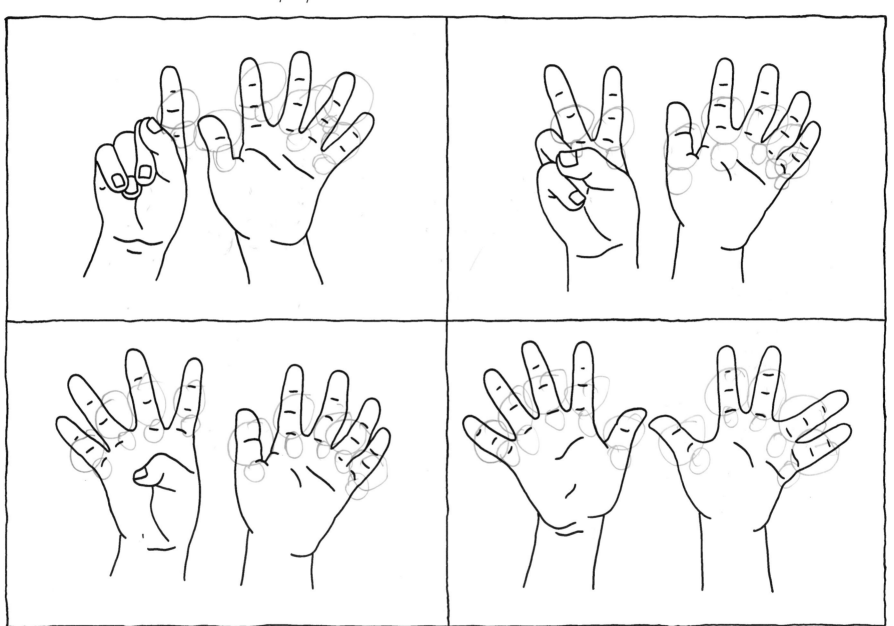

Say, *Look at the first hand. Count the fingers that are standing up.*
Now draw a ring on each finger that is standing up.
Repeat for the other pairs of hands.

Count the funny creatures.

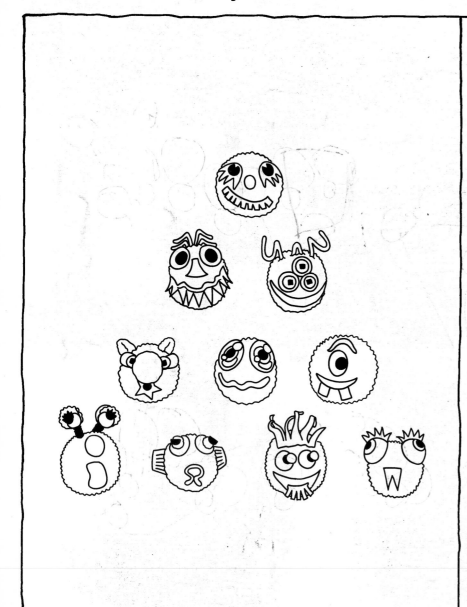

Draw ten funny creatures.

Say, *Count the funny creatures. How many are there?*
Now draw the same number of funny creatures or smiley faces on the other side of the page.

Where is Wilbur?

Name ....

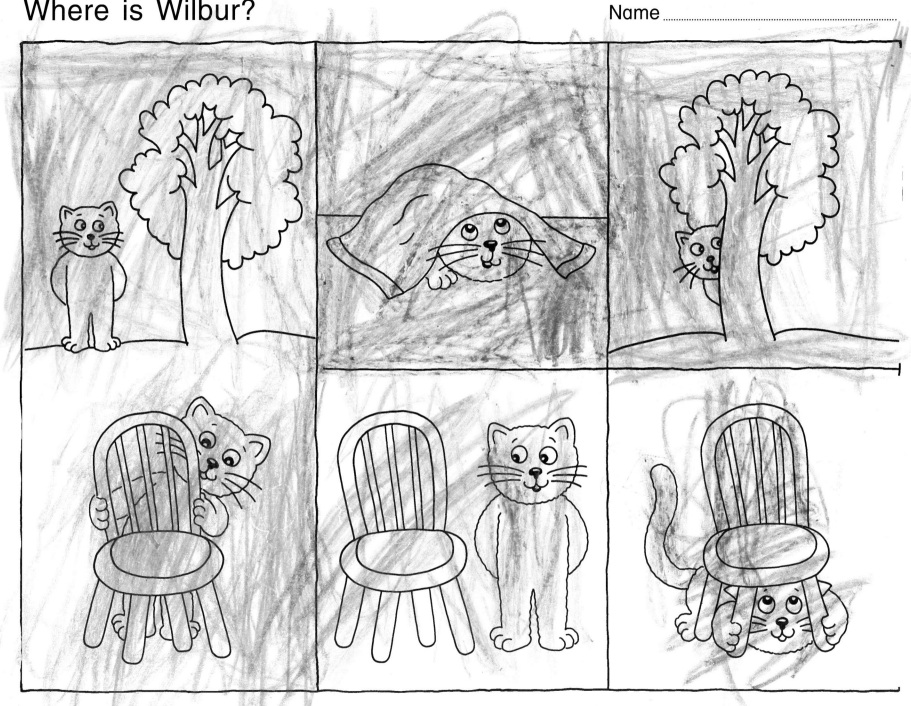

Say, **Find two pictures where Wilbur the cat is behind something. Color those pictures red.**
Find two pictures where Wilbur is underneath something. Color those pictures blue.
Now find two pictures where Wilbur is beside (next to) something. Color those pictures green.

Topic 3   13

Draw a 📖 **on** the table.

Draw an 🍳 **in** the pan.

Draw a 🥛 **next to** the milk.

MILK

Draw 🧀 **under** the table.

Have children look at the first example, then read the instructions with them.
Say, **What do you need to draw? Where are you going to draw it?**
Repeat for the other examples.

Color the shapes that match.

Name _____

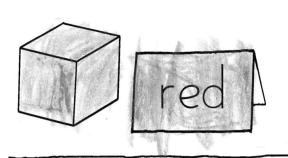   yellow

 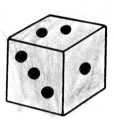 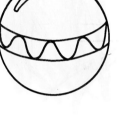 

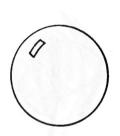  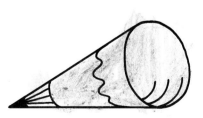

Hold up a classroom cube and say, **This box shape is called a cube. Find the cube at the top of the page.**
Color it red. Now color all the other cubes red.
Repeat for the cones and then the ball shapes (spheres).

Name Janise

Which shapes can **roll**?

    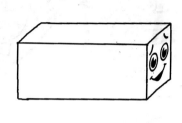

Which shapes can **stack**?

Which shape can **roll** and **stack**?

  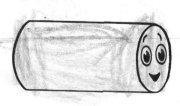  

Revisit *Shape Up!* and discuss what each 3-D character was able to do. Have 3-D classroom shapes available.
Then ask children to look at the first row on this page. Read the question to them and help them decide which pictures they need to color.
Repeat for the other rows of shapes.

Color patterns.

Have the children look at the row of birds and think of some different kinds of patterns they could color. For example, red, blue; red, blue, green; red, blue, blue; and so on. After they have colored the birds, ask, **What kind of pattern did you color?**
Could you have colored a different pattern? What would it look like? Repeat for the other rows.

Color patterns.

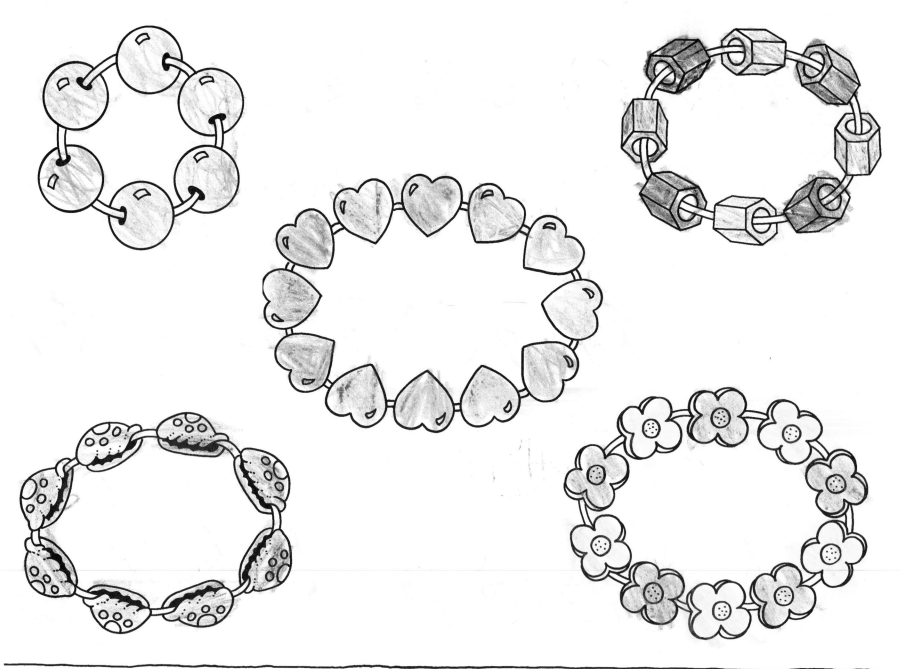

Ask the children to look at the bracelets and describe what they see. Point to the first bracelet and ask how many beads it has. Ask, **What are some ways you could color this bracelet to make a pattern?** After they have colored their pattern, ask them to describe it. Repeat for the other bracelets.

Count. Color to show the number.

1

2

3

4

5

Read the instructions with the children. Then ask them to point to and say the number in the first box, and color that number of worms.
Ask, **How many worms are in the row? How many did you color? How many have not been colored?**
Repeat for the other rows.

Count. Draw more to show the number.

Name ....

5	
3	
4	
2	

Ask the children to tell you what they think they need to do on this page.
Then have them draw items so that the quantity matches the numeral.
Extension: Have children color the items in each row: red for the items printed on the page, and blue for the items they drew.

Show the number of .

five

two

three

four

Ask the children what they think they need to do on this page. Point to "five" and ask, **How many fish are you going to draw in this tank? Draw them.** Repeat with the other tanks.
Extra help: Have children place counters on each tank before drawing the fish.

Color each group that shows the number.

4

5

3

Ask the children to look at the first row and tell you what they see at the beginning of the row.
Say, **Point to each box that shows that many creatures. Color the groups that show four.** Repeat for the other rows.
Extra help: Have children cover the rows they are not using.

Write the numbers.

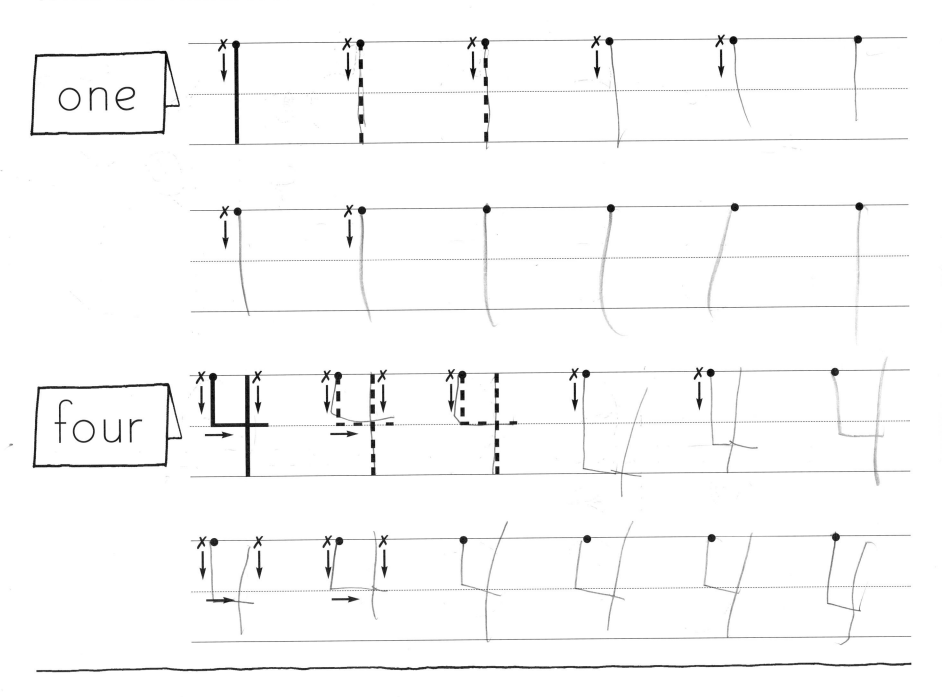

one

four

Extra help: Display the *Writing Numerals 0–5* song poster and encourage children to chant the number rhymes as they write each numeral.

Write the numbers.

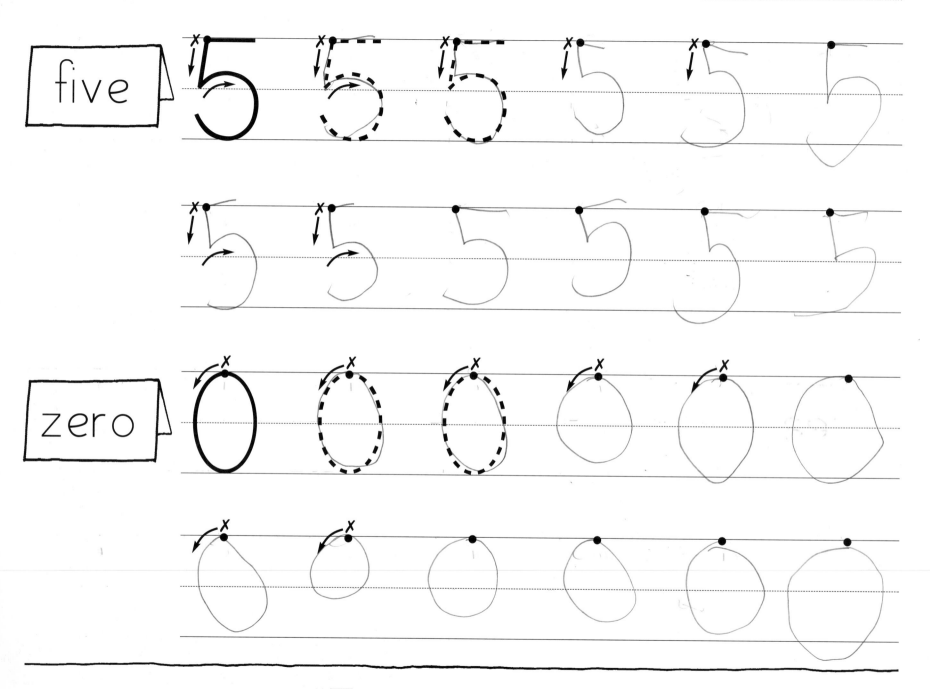

five

zero

Extra help: Display the *Writing Numerals 0–5* song poster and encourage children to chant the number rhymes as they write each numeral.

Write the numbers.

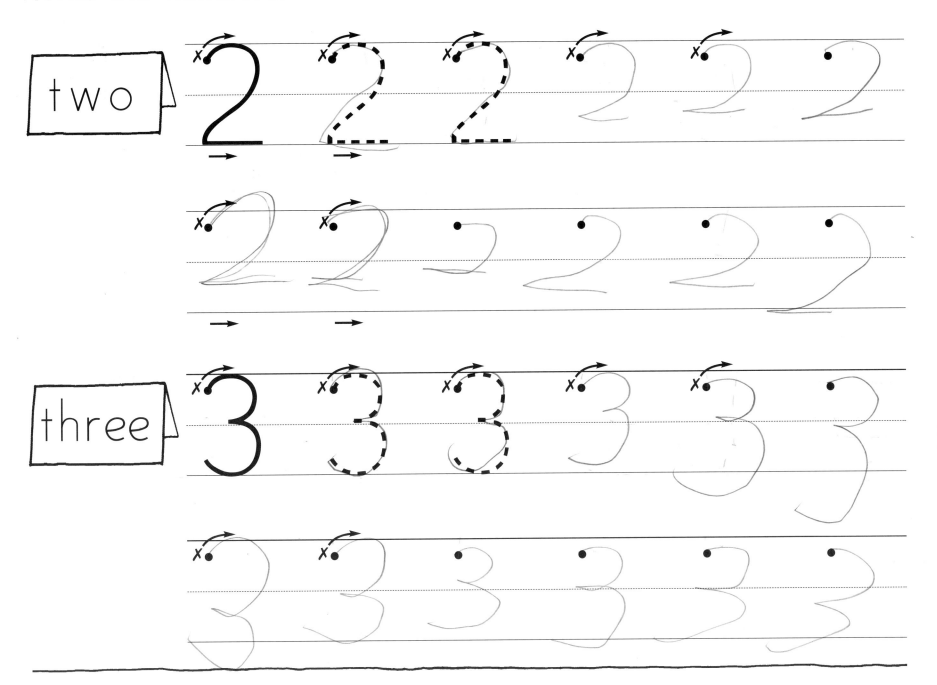

two

three

Extra help: Display the *Writing Numerals 0–5* song poster and encourage children to chant the number rhymes as they write each numeral.

Show the numbers.

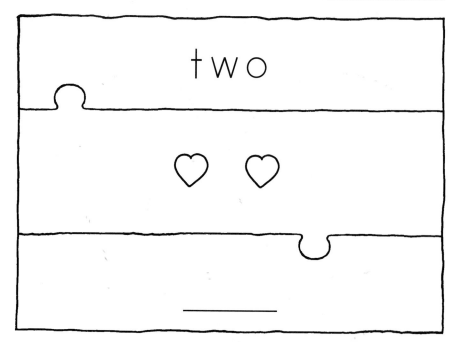

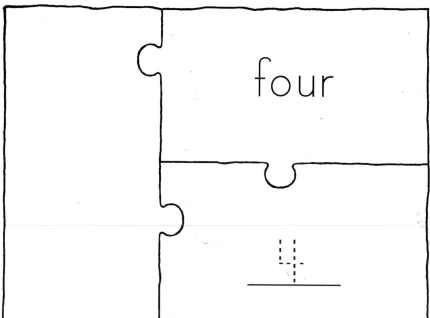

Display *Discussion Book* pages 18–19. Then ask the children to look at the jigsaw pieces on this page and tell you what they need to draw or write on the puzzles. For the first example they write the numeral 5. In other examples they write the numeral and draw hearts. Have them complete the other three puzzles.

Color 3 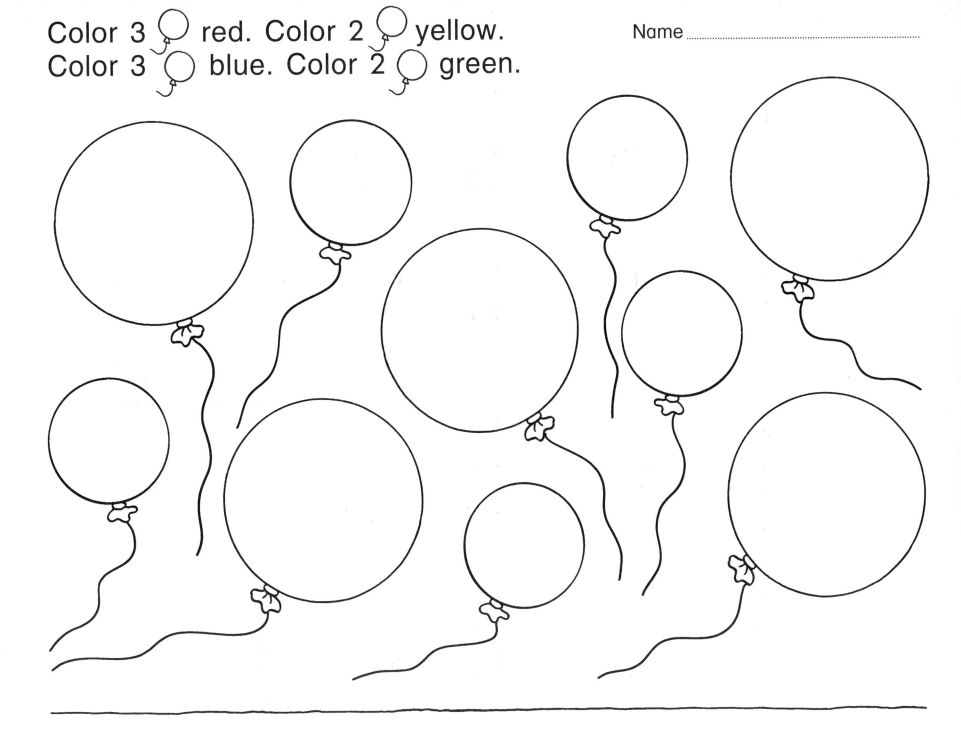 red. Color 2 yellow.
Color 3 blue. Color 2 green.

Read the instructions to the children and allow time for them to color the balloons.
Ask, *How many big balloons are there? How many small balloons? How many balloons are there in all?*

Count. Color to show the number.

Name _____

5

2

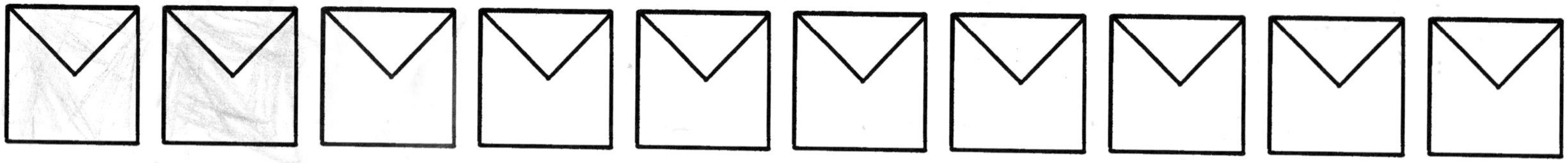

3

2

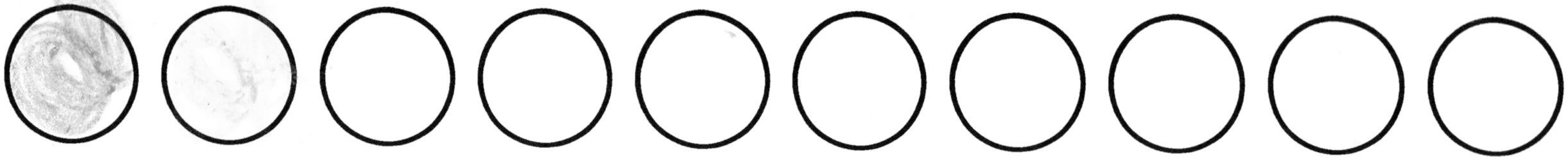

4

Read the instructions with the children. Then ask them to color that number of shells.
Ask, *How many shells are in the row? How many did you color? How many have not been colored?*
Repeat for the other rows.

Match the shapes.

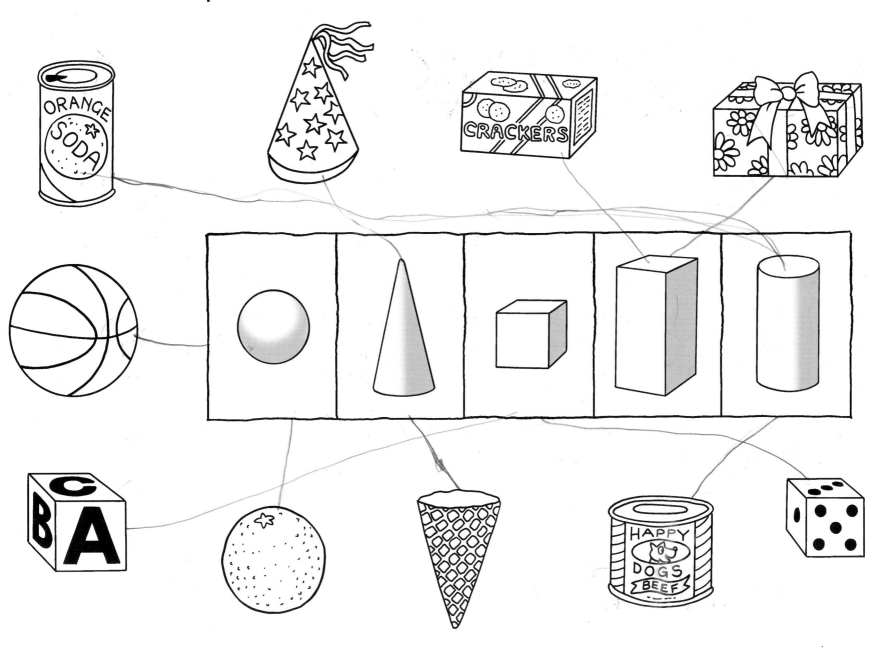

Ask children to look at the 3-D shapes in the center of the page. Ask them to point to the sphere, then say, **Color the ball shape red.**
What other things on the page are shaped like a ball? (Basketball and orange) **Color them red.** Repeat for the other shapes
(cone, cube, rectangular prism/box shape, cylinder). Children can use a different color for each kind of 3-D shape.

What comes next?

Name ....

Make a new pattern.

Have children look at the row of faces and describe the pattern.
Ask, **What kind of face will come next in the pattern? Draw it.** Repeat for the craft sticks and the balls and bats.
Have children draw a pattern with items of their own choice in the space provided.

Draw

three

five

two

four

nine

seven

Say, **Let's read the number words together. What was the first number word we read?** (Three) **Now draw that number of nuts in the bag.**
Repeat with the other examples.
Extension: On a separate piece of paper, have children draw different bags of nuts and write matching number labels.

Count and write.

3

2

5

4

1

0

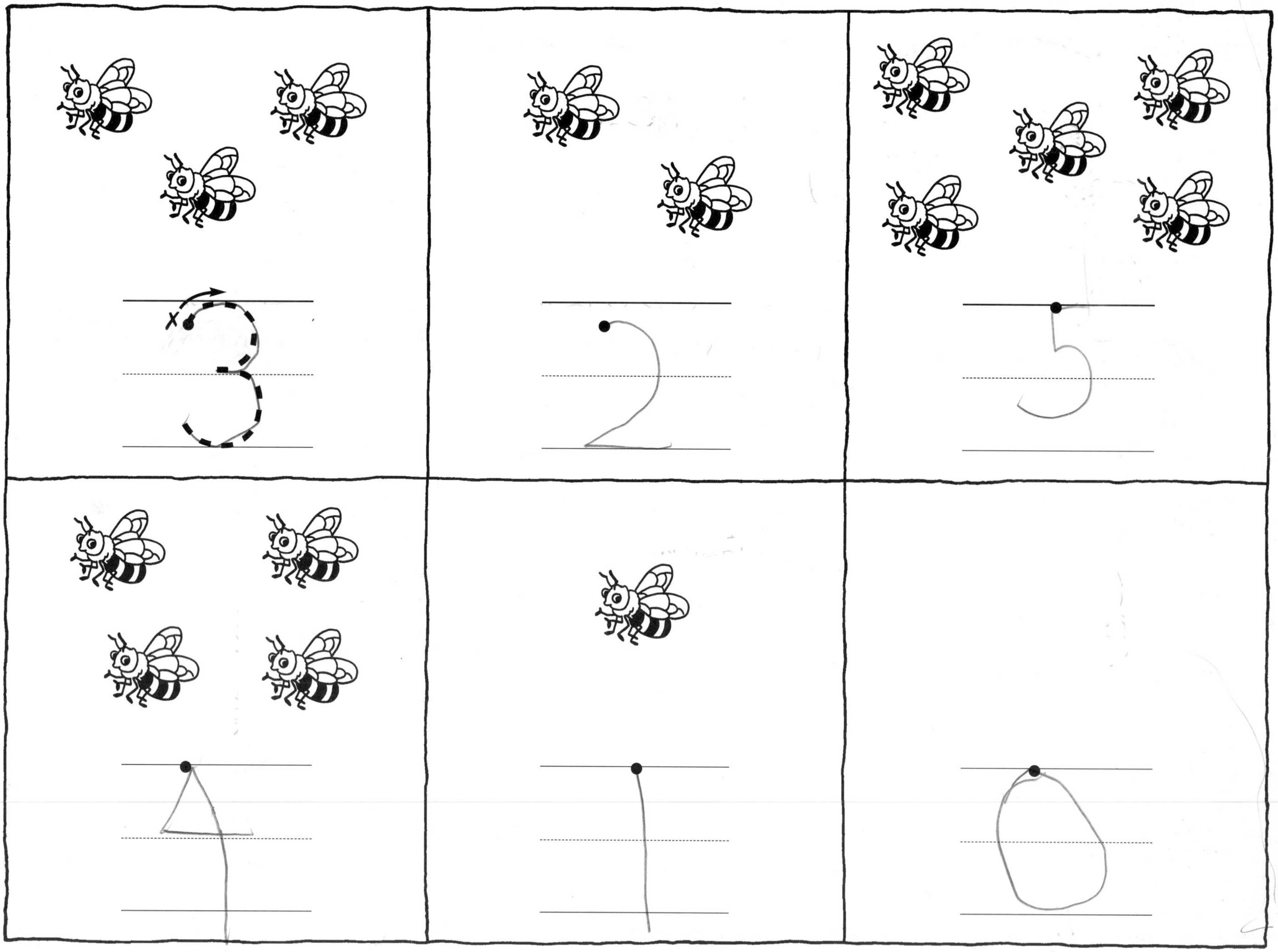

Ask the children to look at the first example and tell you what they think they need to do.
Before they begin to count bees and write the matching numeral, point to the box with no bees.
Ask, **What number will you write to show that there are no bees?** (Zero)

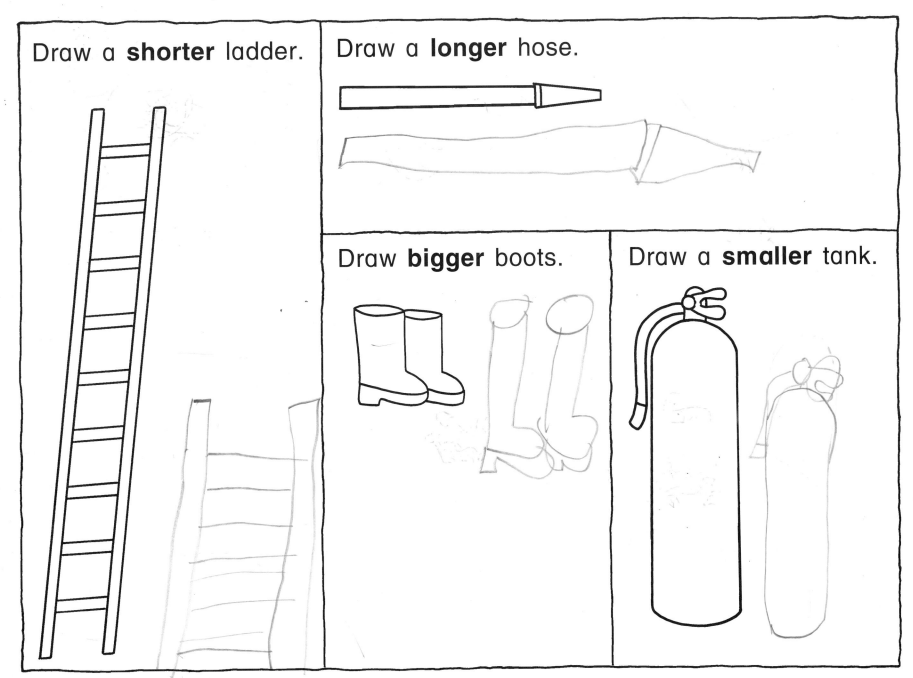

Draw a **shorter** ladder.

Draw a **longer** hose.

Draw **bigger** boots.

Draw a **smaller** tank.

Display *Discussion Book* pages 24–25. Then have the children look at the ladder on this page.
Say, *Next to this long (tall) ladder, draw a ladder that is shorter.* Repeat for the other fire-station items.
Extension: Give children two separate sheets of paper, and have them draw long fire-station items on one sheet and short items on the other.

Draw the next picture.

Name _____

Display *Discussion Book* pages 24–25. Then have the children look at the row of pails on this page.
Ask, ***What do you notice about these pails? What order are they in? What will come next? Draw it.*** Repeat for the flags.
Extension: Give children a separate sheet of paper and have them repeat the activity by drawing an item of their choice in three different sizes.

Color the things that will grow bigger.
Cross out the things that cannot change.

Display *Ben's Beans* pages 2–3. Then say, **Some of the things on your page can change by growing bigger, and some cannot.**
Color the things that will grow longer, taller, or bigger. Cross out the things that cannot change.
Extension: Have children use the space on the page to draw something that can grow.

Color the things that may get smaller.
Cross out the things that cannot change.

Have the children look at the items on this page. Ask, **Which things do you think could change? How might they get smaller or shorter? Which of these things cannot change? Why not?**

Color the group that has **more.**

Ask children to find the two groups of sheep. Say, *How many sheep are in the group that is sleeping? How many are in the group that is running? Which group has more? Which group has fewer? Color the group that has more.* Repeat for the other examples.

Ring the group that has **fewer**.

Ask children to point to the two groups of airplanes. Say, *How many airplanes are gray? How many airplanes are white?*
Which group has more? Which has fewer? Draw a ring around the group that has fewer.
Repeat for the other examples.

Write the numbers.

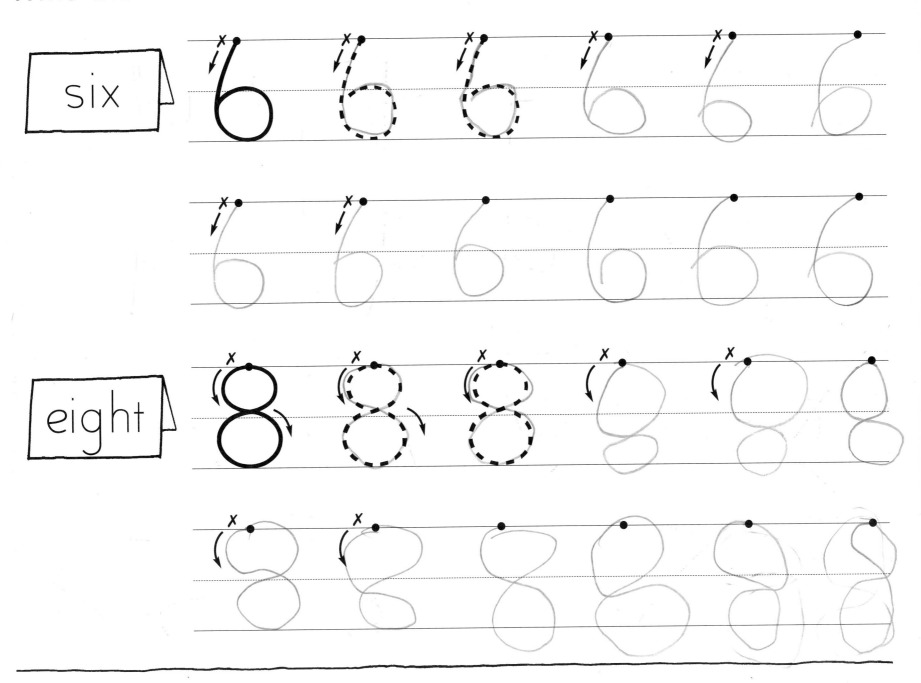

six

eight

Extra help: Display the *Writing Numerals 6–10* song poster and encourage
children to chant the number rhymes as they write each numeral.

Write the numbers.

nine

seven

Extra help: Display the *Writing Numerals 6–10* song poster and encourage children to chant the number rhymes as they write each numeral.

Count. Color to show the number.

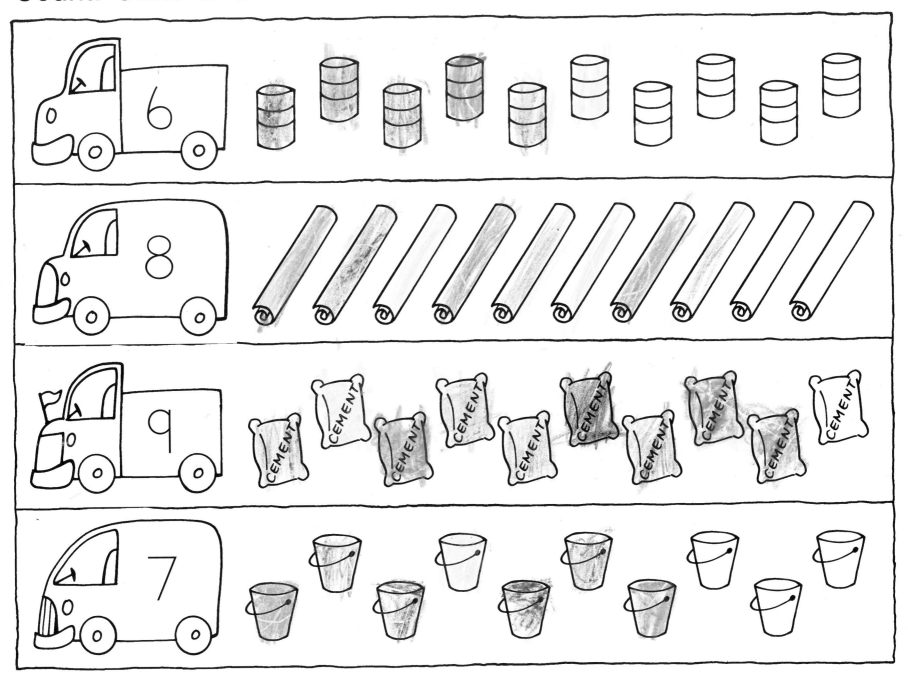

Read the instructions with the children. Then have the children look at the first row. Say, **Read the number on the truck.**
Now color that many barrels. How many did you color? How many have not been colored?
Repeat for the other rows.

Topic 6   41

Count and write.

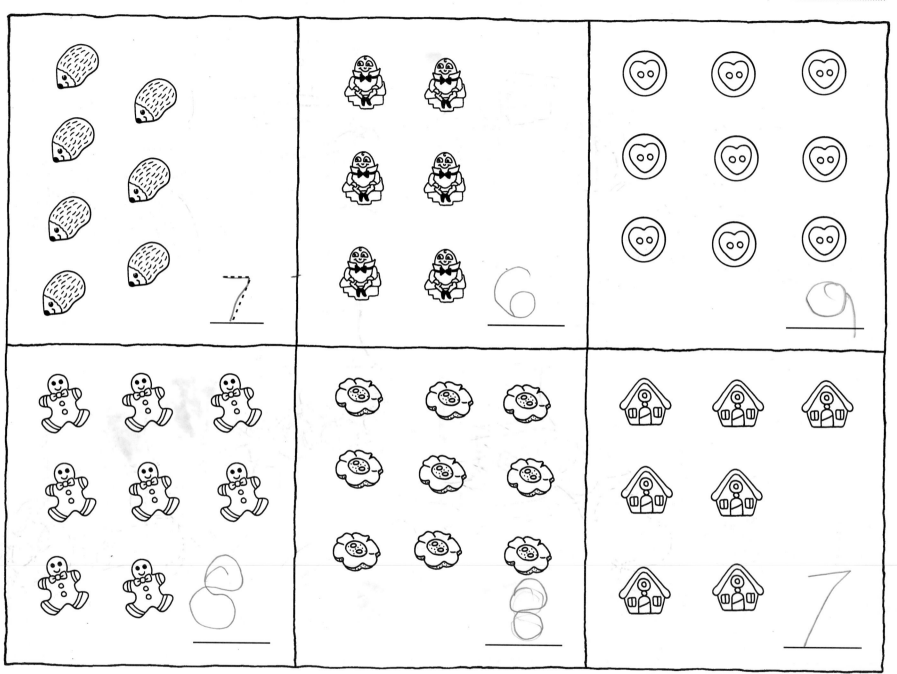

Display the two *Writing Numerals* posters for the children's reference.
Ask the children to look at the first example on this page and tell you what they think they need to do.
Extra help: Children could place a counter on each item to help them keep track as they count.

Count and write.

Display *Discussion Book* pages 28–29, then ask the children to look at the trucks on this page. Ask children to count the number of things on each truck and write the number on the flag. Then ask, **Can you find any trucks carrying the same number of things? Use the same crayon to color them.** Have the children use a different color for each pair of matching trucks.

Draw dots to show **six**.

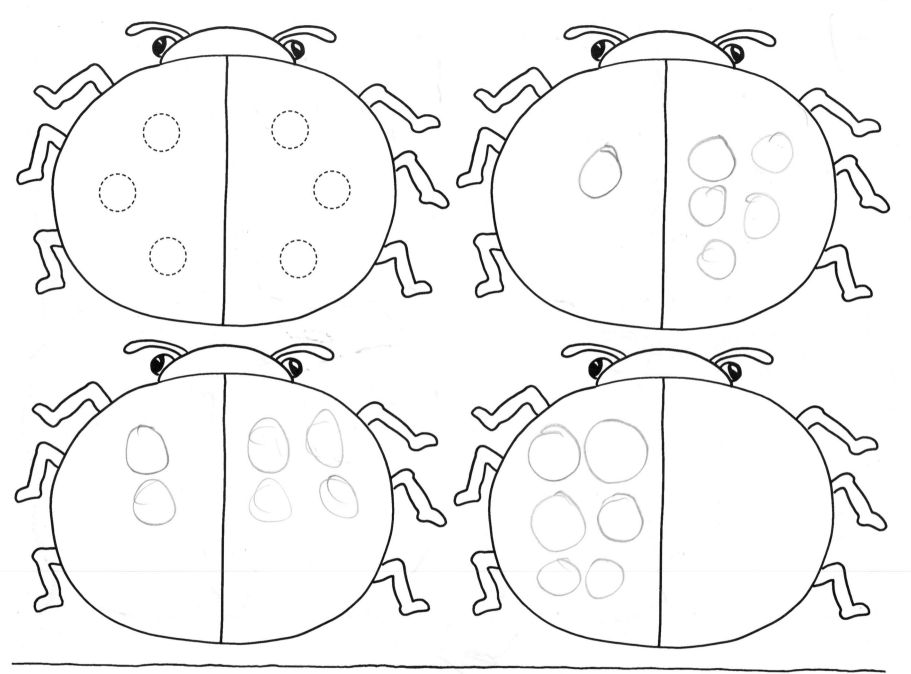

Have children look at the first bug. Say, **Count the dots on one side of the bug. Now count the dots on the other side of the bug. How many dots are there in all? Now draw six dots on each of the other bugs. Make each bug look different.**
Extra help: Before children start drawing dots, provide them with six counters to model their different groups of six.

Draw dots to show **seven**.

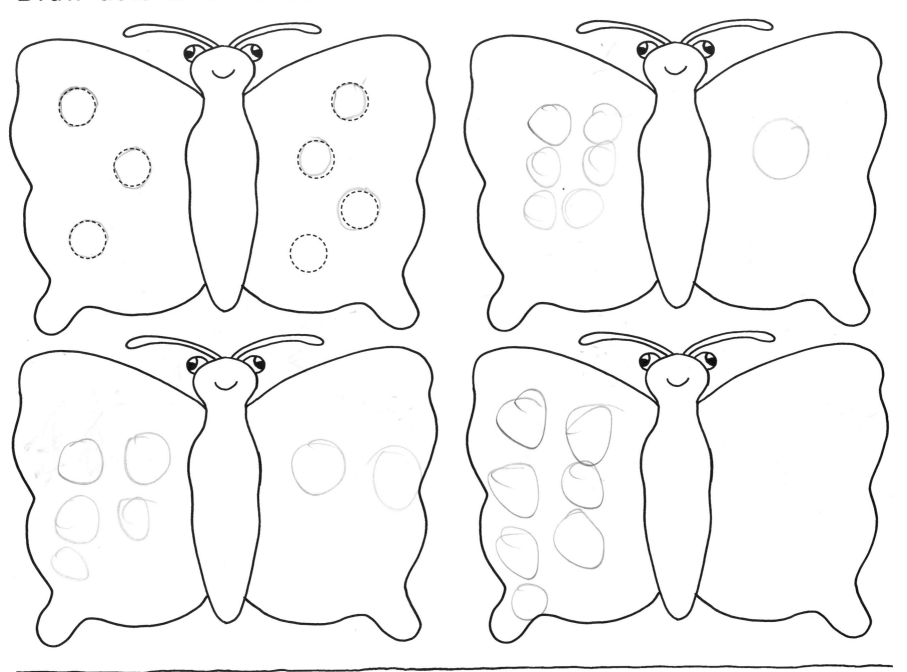

Have children look at the first butterfly. Say, *Count the dots on one side of the butterfly. Now count the dots on the other side of the butterfly. How many dots are there in all? Now draw seven dots on each of the other butterflies. Make each butterfly look different.*
Extra help: Before children start drawing dots, provide them with seven counters to model their different groups of seven.

Draw dots to show **eight**.

Name _____

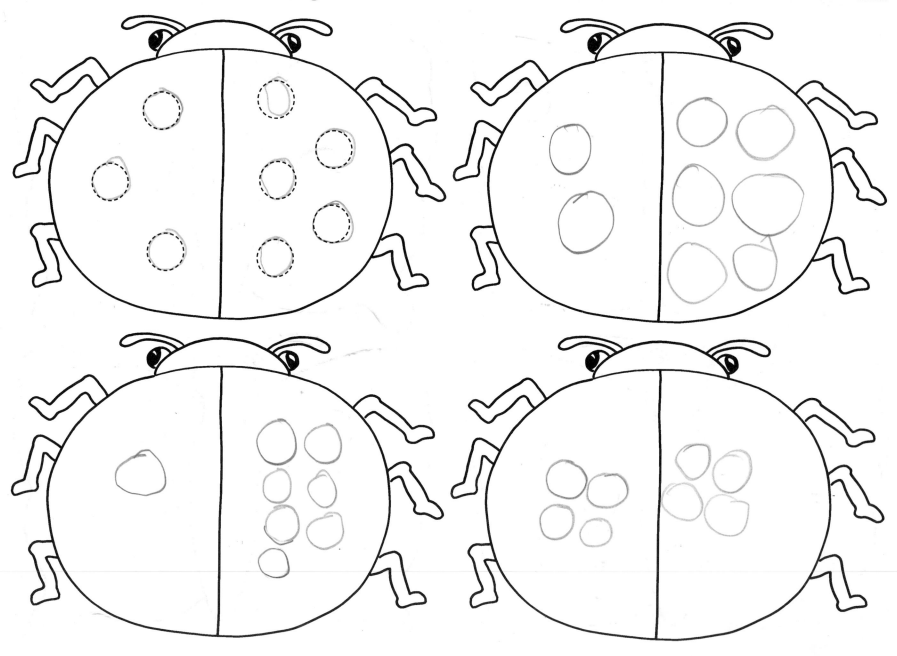

Have children look at the first bug. Say, *Count the dots on one side of the bug. Now count the dots on the other side of the bug.*
How many dots are there in all? Now draw eight dots on each of the other bugs. Make each bug look different.
Extra help: Before children start drawing dots, provide them with eight counters to model their different groups of eight.

Draw dots to show **nine**.

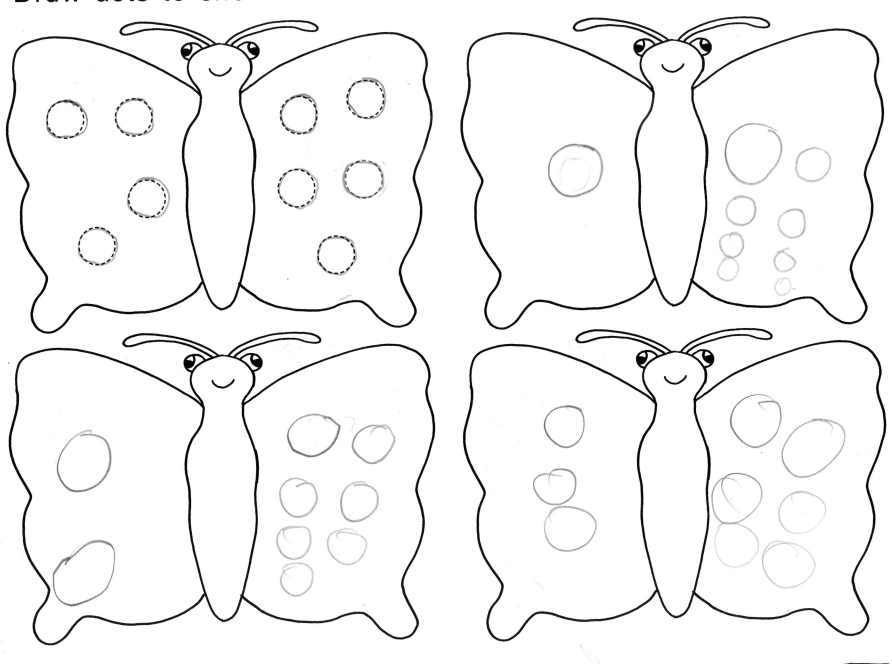

Have children look at the first butterfly. Say, *Count the dots on one side of the butterfly. Now count the dots on the other side of the butterfly. How many dots are there in all? Now draw nine dots on each of the other butterflies. Make each butterfly look different.*
Extra help: Before children start drawing dots, provide them with nine counters to model their different groups of nine.

Count the 2 groups.
Write the number in all.

Name _____

__7__ cookies in all

__0__ cookies in all

__6__ cookies in all

__7__ cookies in all

Have the children look at the butterfly cookies. Ask, *How many are on the tray? How many are beside the tray?*
Make each group a different color. Then write how many butterfly cookies there are altogether.
Repeat the questioning for each of the other kinds of cookies.

Color the circle flowers.
Draw a circle flower for each stem.

_____ circle flowers in all

Have the children look at the printed flowers. Ask, *What shapes are the flowers made from?* (Circle, square, triangle) *Color the circle flowers.*
Now draw a circle flower on each empty stem. Color your flowers.
Extension: When children have finished coloring and drawing, have them count the circle flowers, and write how many there are altogether.

Topic 7   49

Trace and count.

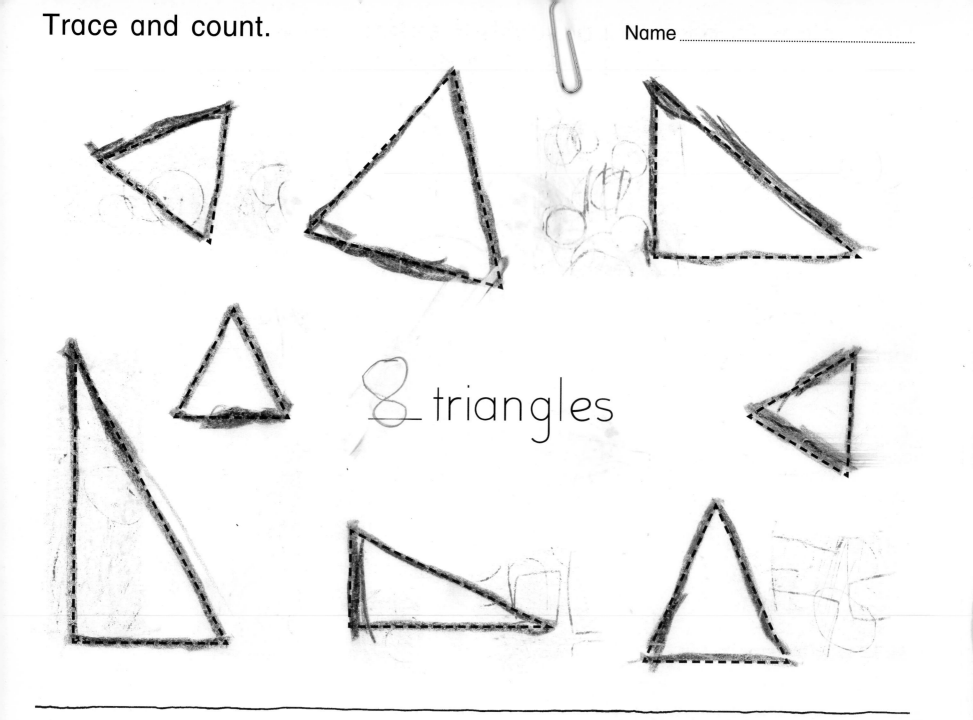

8 triangles

Ask, **What shape do you see on this page? Do all the triangles look the same?**
What is the same about them? What is different? Have the children trace over each triangle,
first with a finger, then with a crayon. Then have them count and write the number of triangles.

Use yellow to color shapes with **3 sides**.
Use red to color shapes with **4 sides**.

Ask, **What shapes are on this page? Which of them have straight sides? Which do not have straight sides?**
What do we call a shape that has three straight sides? (Triangle) **Color the triangles yellow.**
What do we call a shape that has four straight sides? (Rectangle, or square) **Color all these shapes red.**

Topic 7   51

Color the squares.
Draw a new square.

Name _____

Have the children look at all the shapes and draw a dot on those they think are squares.
Ask, ***What other shape can you see that has four sides and four corners? What is its name?*** (Rectangle)
Have the children color the squares and draw their own square in the space provided.

Keep the patterns going.

Make your own shape pattern.

Have the children look at the first row. Ask, **What pattern can you see?** (Circle, rectangle, circle …)
Say, **Now draw more shapes to make the pattern continue (go on and on).** Repeat for the other three patterns.
Extension: Have children draw two shapes and repeat them to make a pattern in the space provided.

Color the shapes. Count them.

Name _____

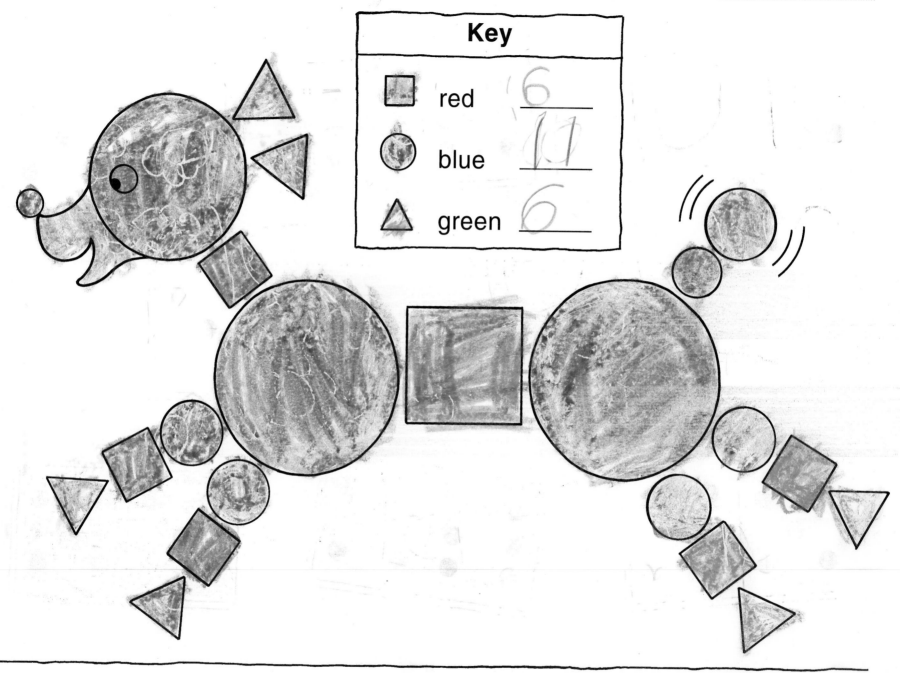

Key

▢	red	6
◯	blue	11
△	green	6

Show children the key and have them color the small square red. Say, **_Now find all the other squares. Color them red._**
Repeat for the circles and the triangles. Then have the children count how many of each shape are in the dog picture and write the number.
Extra help: Have children place a counter on each (square) before counting and writing the number.

Color the ▢▢ that show ten.

Ask the children to look at the first domino. Ask, **How many dots are on the left side of the domino?**
How many dots on the right side? How many are there altogether? Color the domino.
Repeat for the other examples.

Write the numbers.

ten

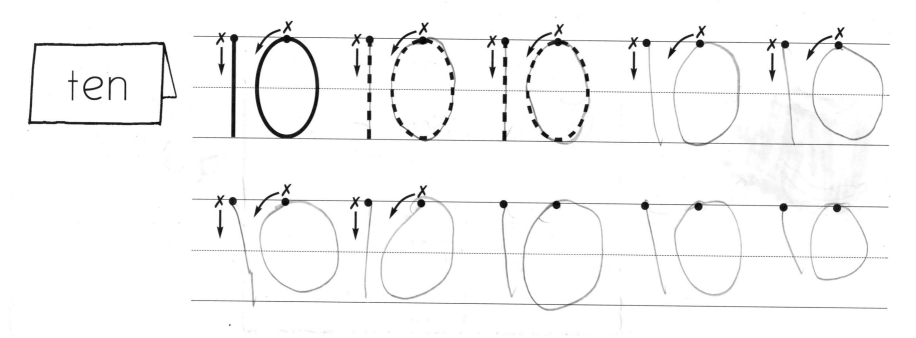

Write the number word.

ten ten ten ten ten

Display the *Writing Numerals 6–10* song poster and read the number rhyme for ten with the children. Have them practice writing the numeral and the number word for ten.

Count. Draw more to show the number.

10

10

10

10

Ask the children to look at the first example. Say, *How many bananas do you see? Draw more bananas to make 10. Trace over the number.*
Extension: Have the children use two colors to color the items in each box: red for the items printed on the page, and blue for the items they drew.
Ask, *How many (bananas) were already there? How many did you draw? How many are there in all?* Repeat for the other examples.

Draw ✿.

4	✿	✿	✿	✿						
5										
6										
7										
8										
9										
10										

Have the children look at the first row. Say, **Count the flowers. How many are there?**
Now draw flowers in each row to match the number.
Extra help: Have children start by placing counters on the boxes in a row to model the correct number for that row.

Join the numbers in order.

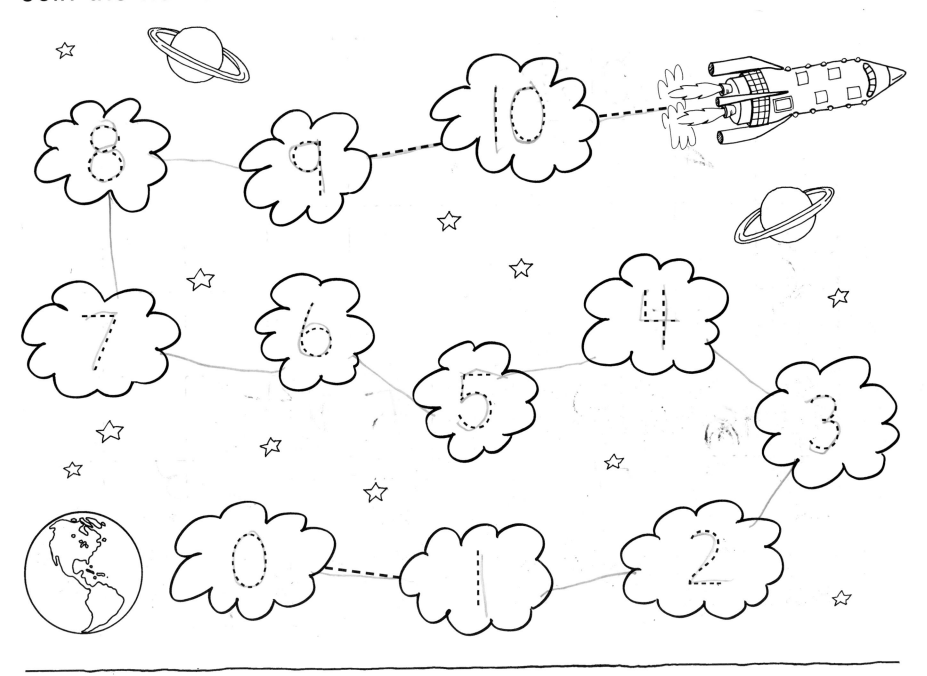

Read the instruction with the children. Ask, **Where would you like to start?** (Zero or ten) **Why?** Have them trace over each number and then draw the path.
Extra help: Guide the children to follow the path between the numbers with their finger.
Extension: Ask those children who counted forward from zero to now count backward from ten, and vice versa.

Write the missing numbers.

Name ....

Ask the children to look at the row of houses.
Ask, **What numbers are missing? Fill in the missing numbers.**
Repeat for the line of boats.

Show the numbers.

Have the children look at the jigsaw pieces that make up the first puzzle.
Ask, *What do you need to draw or write on the puzzle?* (The numeral 7)
Repeat the questioning and have the children complete the other three puzzles.

Count the 2 groups.
Write the number in all.

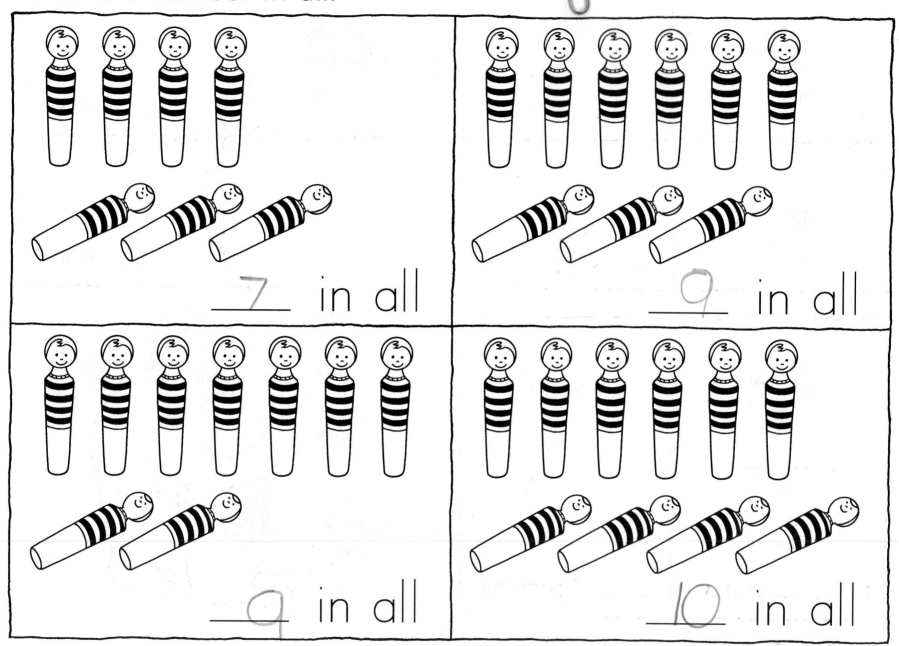

_7_ in all

_9_ in all

_9_ in all

_10_ in all

Have the children look at the pin people in the first box.
Ask, *How many are standing up? How many are lying down? Make each group a different color.*
Then write how many pin people there are altogether. Repeat for the other three boxes.

Write the numbers.

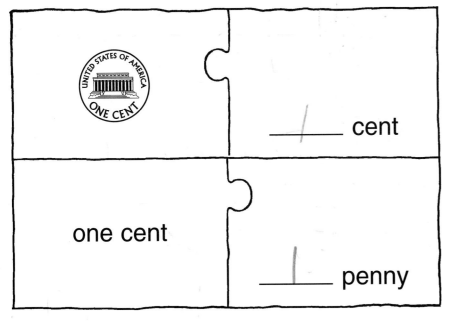

_____ 1 cent

one cent

_____ 1 penny

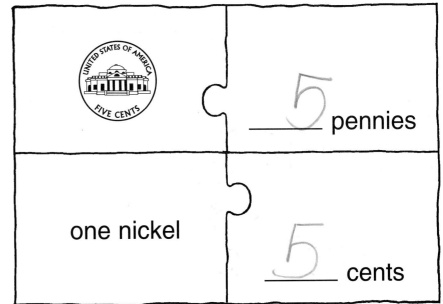

_____ 5 pennies

one nickel

_____ 5 cents

one dime

_____ 10 cents

_____ 10 pennies

Display *Discussion Book* page 44. Then have the children look at the first puzzle on this page. Ask, **What do we call this coin?**
What color is a penny? Color it brown. How many cents is a penny worth? Write the missing numbers.
Repeat the activity for the nickel and the dime.

Count the 2 groups of pennies.

Name _____

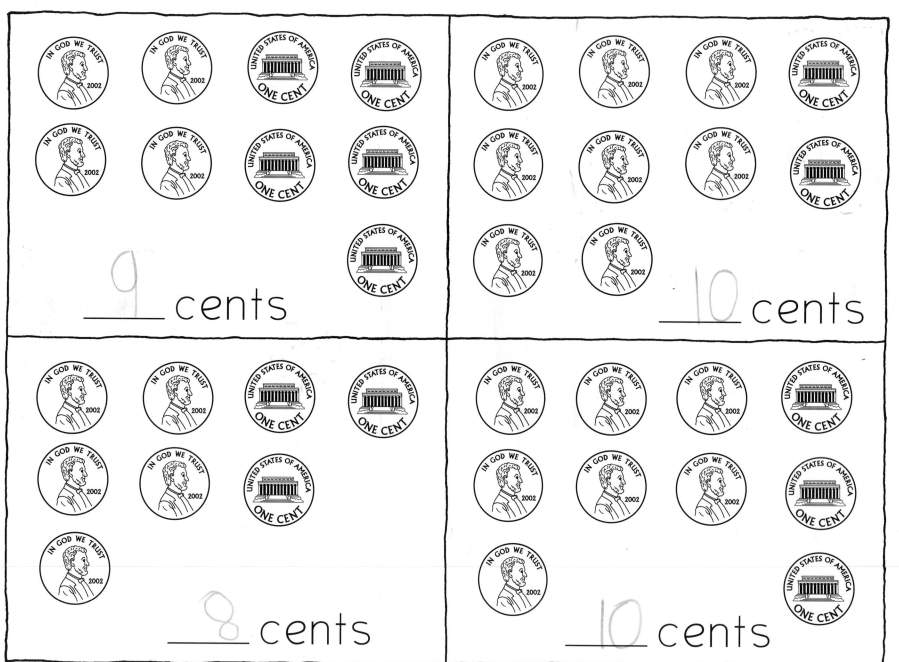

_9_ cents

_10_ cents

_8_ cents

_10_ cents

Have the children look at the coins in the first box. Ask, **What are these coins? How many pennies are showing heads? How many are showing tails? Write the number of cents.** Repeat for the other three boxes.
Extra help: Have children place a real penny on each coin picture to help them count.

Draw **long** legs.　　Draw **short** legs.　　Draw a **longer** snake.

Draw a **shorter** flower.

Read the "puppet" instructions with the children. Then have them draw long legs on the man puppet and short legs on the woman puppet. Say, **_Tell me about the legs you drew._** Repeat for the snake and the flower.

Keep the patterns going.

Make your own shape pattern.

Have the children look at the first row. Ask, **What pattern can you see?** (Square, circle, square …)
Say, **Now draw more shapes to make the pattern continue (go on and on).** Repeat for the other three patterns.
Extension: Have children draw two shapes and repeat them to make a pattern in the space provided.

Count. Draw more to show the number.

Name _____

_9_ nine

_8_ eight

_7_ seven

_6_ six

Ask the children to look at the first example and say, **How many triangles do you see? Draw more triangles to make nine. Trace over the number.**
Extension: Have the children use two colors to color the shapes in each box: red for the shapes printed on the page, and blue for the shapes they drew.
Ask, **How many (triangles) were already there? How many did you draw? How many are there in all?**

Draw 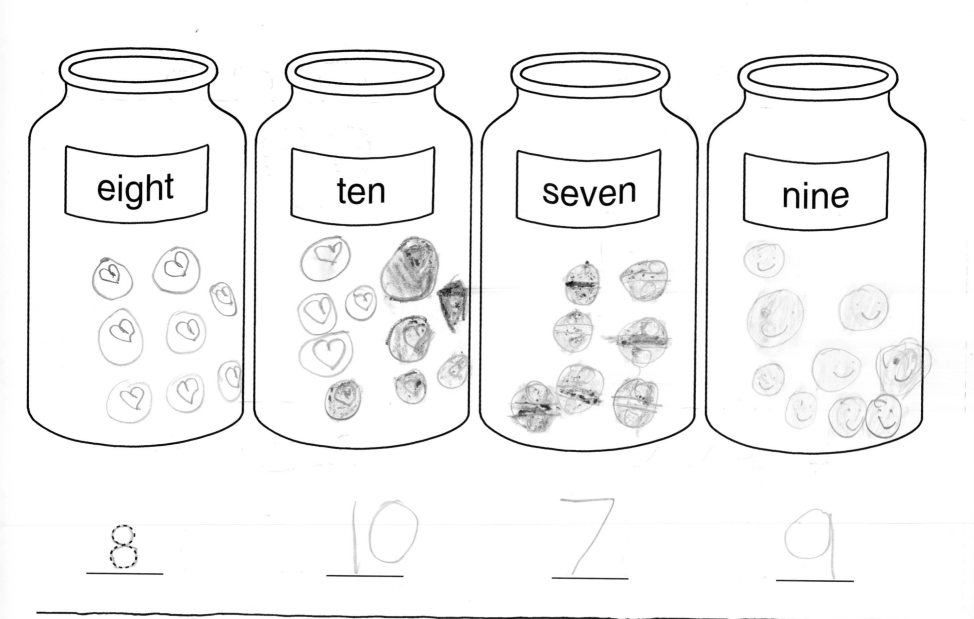 in the jar.
Write the number.

Name ....

eight

ten

seven

nine

8

10

7

9

Have the children read the number word label on the first jar.
Say, **Draw that number of marbles in the jar. Then write the number below the jar.**
Repeat for each of the other jars.

Count the 2 groups.
Write the number in all.

__8__ in all

__7__ in all

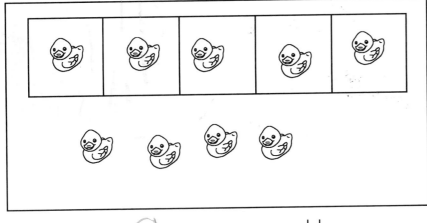

__9__ in all

__10__ in all

Have the children look at the first example. Ask, *How many pigs are on the five frame? How many are below the five frame? Color each group a different color. Then write how many pigs there are altogether.* Repeat for the other examples.

Count the 2 groups.
Write the number in all.

Name _____

_9_ in all

_8_ in all

_10_ in all

_10_ in all

Have the children look at the first example. Ask, *How many tyrannosaurus dinosaurs are there? How many brontosaurus dinosaurs are there? Color each group a different color. Then write how many dinosaurs there are altogether.* Repeat for the other examples.

Color the groups of **eleven**.

Ask children to look at the first group of bugs. Say, **Count the bugs. How many are there?**
If there are eleven, color the picture. Repeat for the other examples.
Extra help: Have children put a dot on each item as they count it.

Color the groups of **twelve**.

Name ....

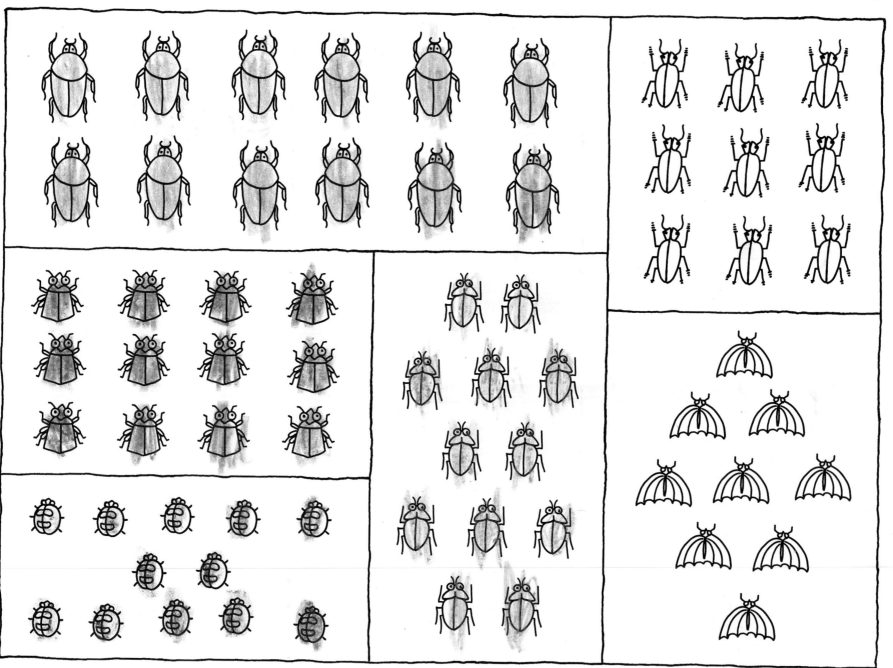

Ask children to look at the first group of bugs. Say, **Count the bugs. How many are there?** **If there are twelve, color the picture.** Repeat for the other examples.
Extension: Have children draw groups of twelve things on a separate sheet of paper.

Count. Draw more to show the number.

Name ....

9

11

12

10

11

12

Ask the children to read the number in the first box. Ask, **What is the number in the box?**
How many moons are there? Draw more moons so that you have nine.
Repeat for the other examples.

Count. Color to show the number.

13

15

12

14

Ask the children to look at the first box. Say, **Read the number. Count the ducks. How many are there?**
How many do you need to color? Repeat for the other examples.
Extension: Have children tell you how many items are colored and how many are not colored.

Count. Color to show the number.

Name

15

14

13

12

11

Ask children to tell you what they think they need to do on this page. When they have finished coloring the nuts to match the numerals, ask, **Do you notice anything special about the way the nuts are colored?** (The number of nuts not colored increases every time.)

Count. Draw more to show the number.

Name _____

10	15	12
11	13	14

Ask the children to look at the first box. Say, **Read the number.**
How many creatures are there? Draw more creatures so that you have ten.
Repeat for the other examples.

Color things that are about the same length as the ▢▢▢▢◗.

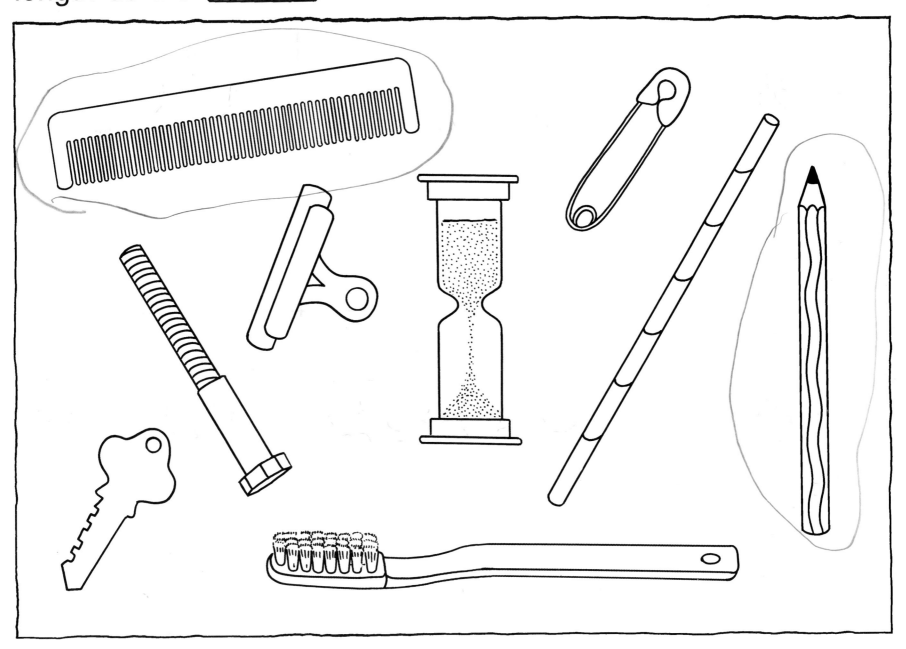

Provide the children with five connecting cubes and ask them to make a measuring rod. Ask, **Which things on this page do you think are about the same length as your measuring rod? Use your measuring rod to check. Color those things red.**
Extension: Have children use different colors to color items that are shorter than the rod and longer than the rod.

Use 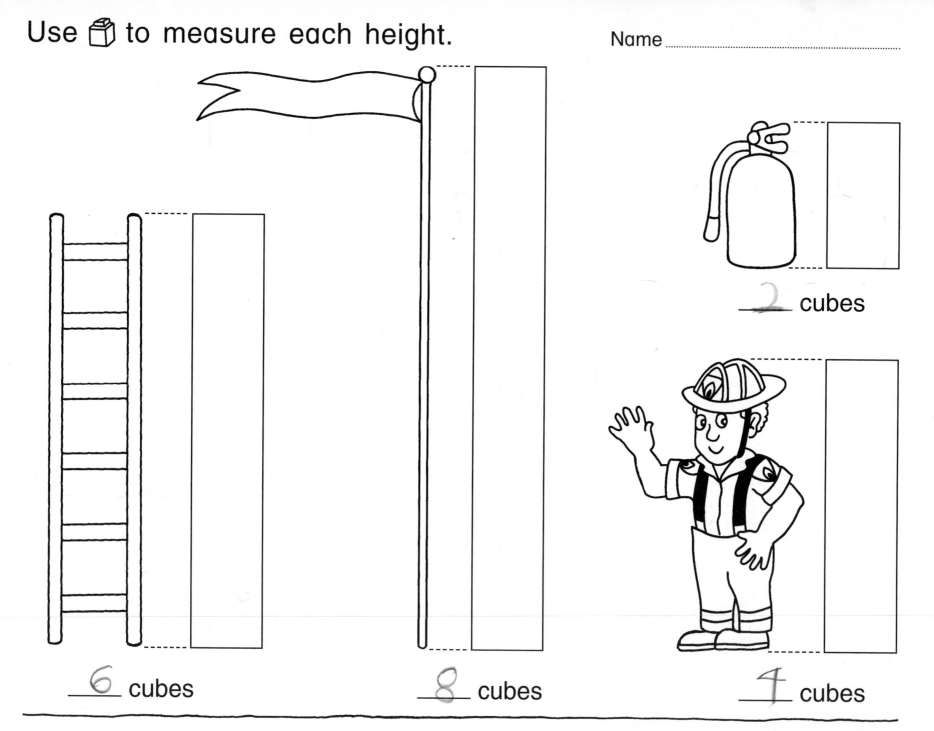 to measure each height.

_6_ cubes

_8_ cubes

_2_ cubes

_4_ cubes

Provide at least eight connecting cubes for each child. Ask children to point to the box next to the ladder.
Say, **_Place connecting cubes side by side along the box. Count the cubes. Write the number._**
Repeat for the other examples.

Count and write.

S	M	T	W	T	F	S

(weather calendar grid with sun, rain, clouds, and snow-covered tree pictures)

Sun — 4

Rain — 1

Snow tree — 0 (circled)

Cloud — 0 (circled)

Have children look at the weather pictures on the calendar. Say, **Look at the sunny day pictures.**
How many can you find? Write the number on the line next to the sun.
Repeat for the other weather pictures.

Show the order of the dinosaur's day.

Name _____

Ask the children to look at the pictures of the dinosaur's day. Say, **What do you think you need to do? There's a clue to help you. What happens first? Trace over the number one.** Then have children write a number in each box to show the order of events.
Extra help: Photocopy the page and have the children cut out separate events. They can put the cards in order and then write the numbers.

Count. Color to show the number.

Name

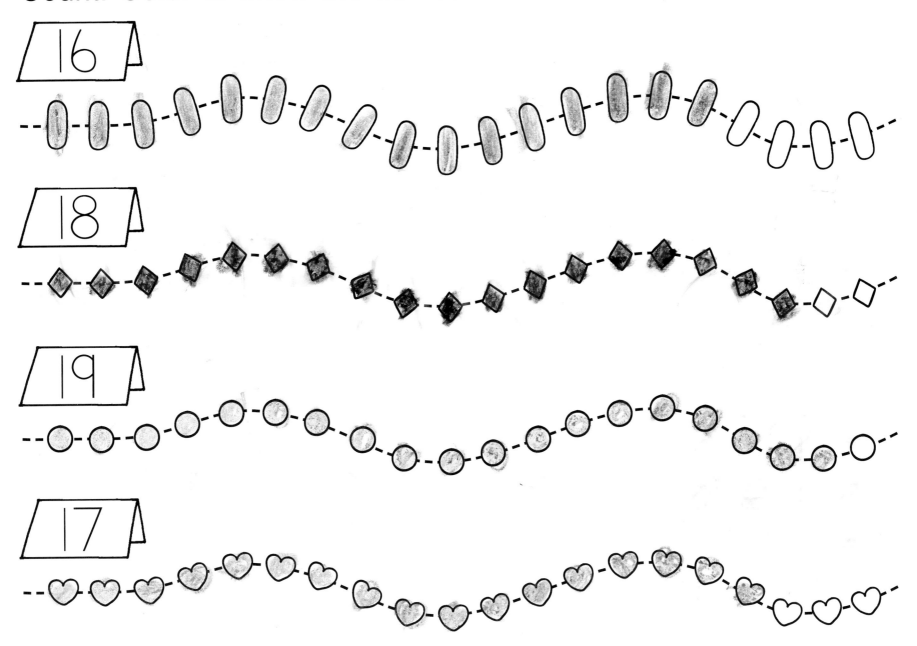

16

18

19

17

Read the instructions with the children. Have them look at the first string of beads.
Say, **Count as you color each bead. Stop when you have colored 16 beads.**
Repeat for the other strings of beads.

Color the groups of **sixteen**.

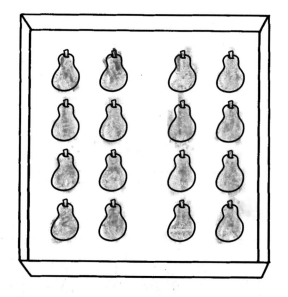

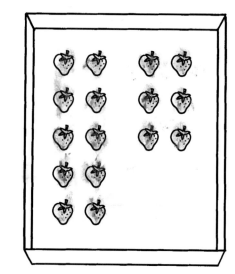

16

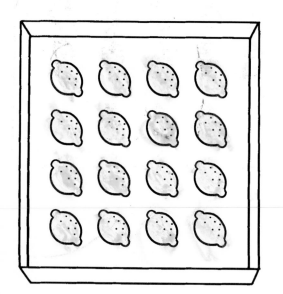

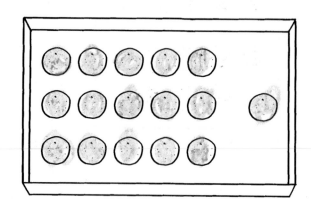

Display *Discussion Book* pages 56–57. Then ask the children to look at this page.
Say, **Count the pears. How many are there? If there are sixteen, color the picture.**
Repeat for the other examples.

Color the groups of **eighteen**.

Name ....

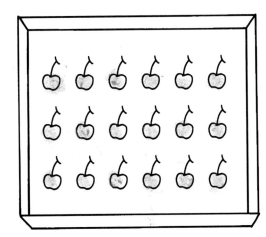

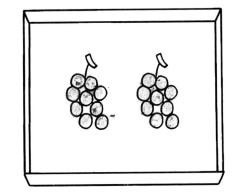

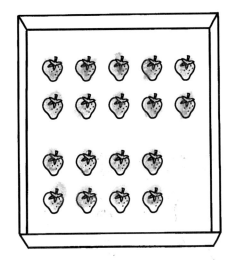

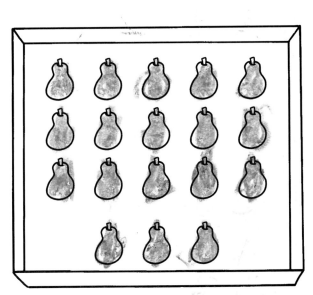

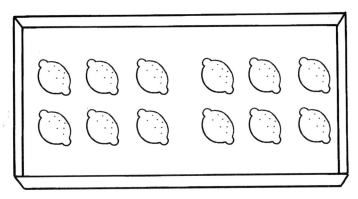

Display *Discussion Book* pages 56–57. Then ask the children what they think they need to do on this page.
Say, ***Count the cherries. How many are there? If there are eighteen, color the picture.***
Repeat for the other examples.

Topic 11   **83**

Color 20 bugs.

Name _____

Read the instruction with the children and have them color twenty of the bugs.
Extension: Have children count all the bugs on the page. Encourage them to count by fives.

Count and color.

Name ....

12	13	14	15	16	17	18	19	20

Have the children look at the "12" column. Ask, **How many bricks do you think there are for this number? Count them and color them red.** **Then choose a different crayon and color the bricks for 13.** Have the children continue the activity, using different colors for each column.
Extension: Choose two numbers on the wall and have children compare the numbers.

Write the missing numbers.

Name _____

1 2 3 4 5 6 7 8 9 10

1 2 3 4 5 6 7 8 9 10

Have the children say the numbers in order from one to ten.
Say, **What numbers are missing? Write them in order on the rocks.**
Extra help: Display *Discussion Book* pages 40–41 for the children's reference.

Count. Draw more to show the number.

Name ....

7 cents in all

8 cents in all

9 cents in all

Have the children look at the first purse in the top row. Say, **Count the pennies in the purse. How many cents is that? Read how many cents you need in all. Now draw more pennies in the second purse so that you have seven cents in all.** Repeat for the other examples.
Extra help: Have children act out each situation with real coins before drawing the missing pennies.

Buy 2 things. Draw the coins.

Name ....

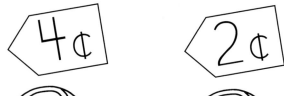  6 ¢

 7 ¢

 8 ¢

 8 ¢

Have the children look at the first row. Ask, *How many pennies do you need to buy one car? Draw the pennies. How many pennies do you need to buy the other car? Draw those pennies. Write how many cents you need in all.* Repeat for the other rows of toys.
Extra help: Before drawing, children could place real coins in one row.

Draw 2 legs on each doll.

Name ....

____8____ legs in all

____6____ legs in all

___10___ legs in all

Have the children look at the paper dolls in the first row. Say, **Draw two legs on each doll.**
Count the dolls. Now count the legs. Write the number of legs in all.
Repeat for the other rows of paper dolls.

Draw equal groups.

Name _____

2 bags of 3

2 bags of 6

3 bags of 4

2 bags of 7

Display *The Squirrels' Store.* Then have children look at the first example on this page. Ask, ***How many nuts are in the first bag? We need to show the same number of nuts in the second bag. How many nuts is that? Draw the nuts so you have two equal groups.*** Repeat with the other examples.
Extension: Give children a separate sheet of paper and have them draw equal groups of other numbers of nuts.

Share the 🐟 onto 2 ⬭.

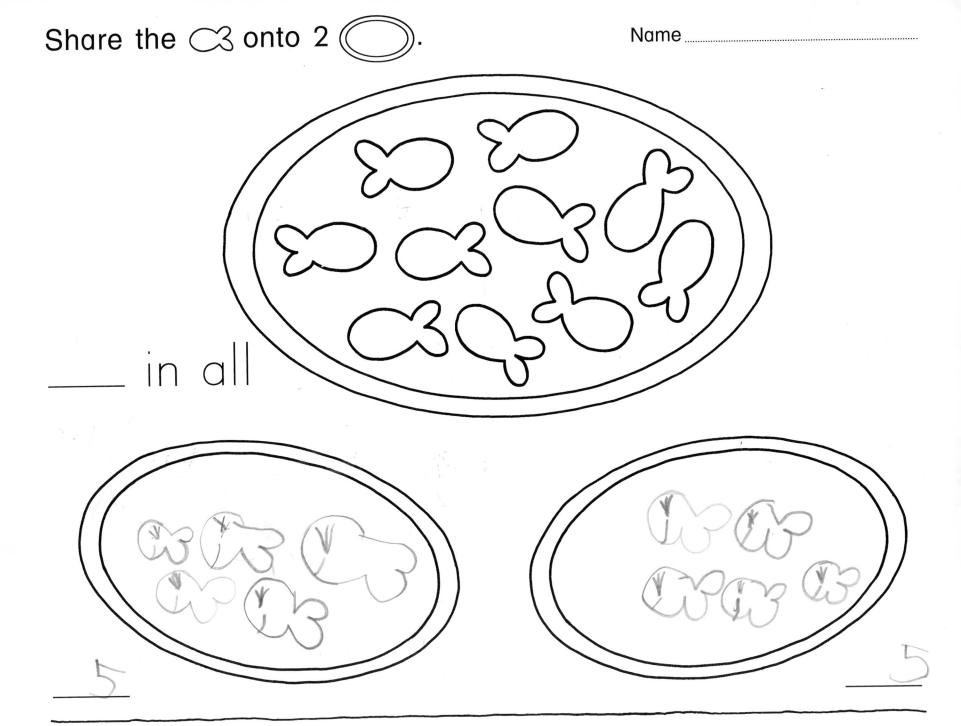

___ in all

5   5

Provide children with counters and have them place a counter on each goldfish.
Say, **Count the counters. Now share them between the two plates. Draw a fish for every counter you shared onto the small plates.**
Extension: Have children write the total number and the number in each share.

Topic 12   91

Which shapes can you fold in half? Color them.

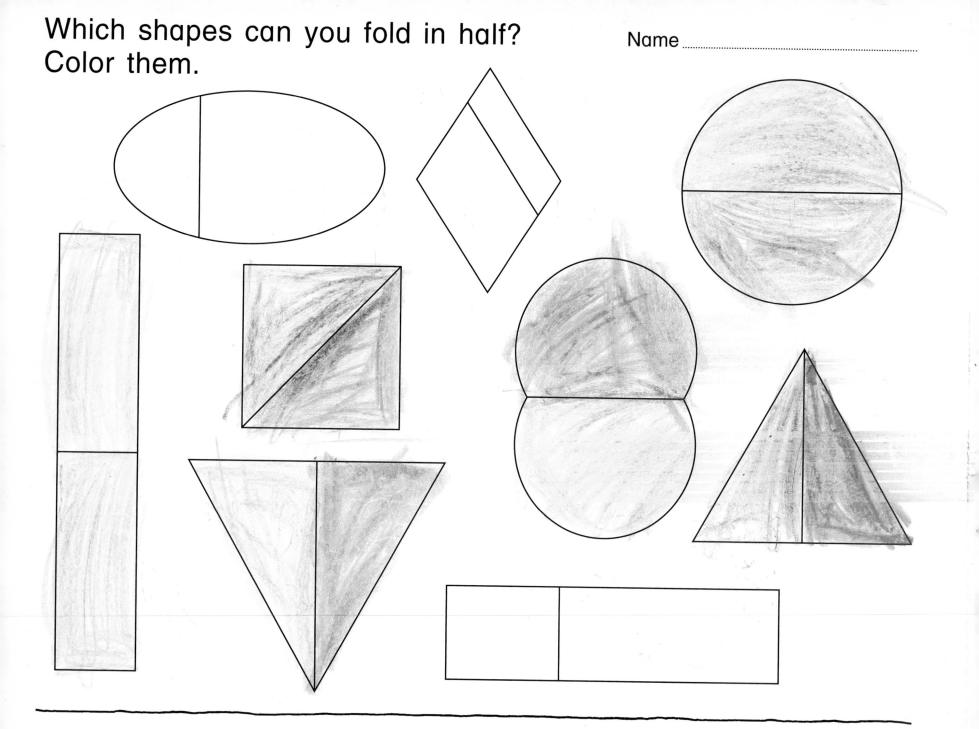

Ask the children to look at all the shapes on the page.
Say, **Imagine each shape is on a separate sheet of paper. Would you be able to fold the shape in half?**
Extra help: Provide a photocopy of the page. Have children cut out each shape and try to fold it in half.

Color silverware on the **left**.

Color silverware on the **right**.

Draw a ⊻ on the **left**.

Draw a ⧄ on the **right**.

Display the *Big Bear Wobble* song poster and review "left" and "right." Then read the first instruction on this page with the children and have them color the forks on the left of the plate. Repeat with the other examples.

Count and write.

Toy	Number
rabbit	3
doll	6
fire truck	2
teddy bear	8
soccer ball	5

Say, *Count the toy rabbits. How many are there?* Have children record the number of rabbits they count in the table. Repeat for the other toys.
Extra help: Have children place counters on each (rabbit), count the counters, and write the number.

Draw equal groups.

2 jars of 6 🥜

3 jars of 4 🥜

Have children look at the top row. Ask, *How many nuts do you need to draw in the first jar? How many nuts do you need to draw in the second jar? Draw two equal groups of nuts.* Repeat the instructions for the jars in the bottom row.
Extension: Have children count the nuts in the top row and then the bottom row. Ask, *What do you notice?*

Join the dots. Start at 1.
Finish at 20.

Name _____

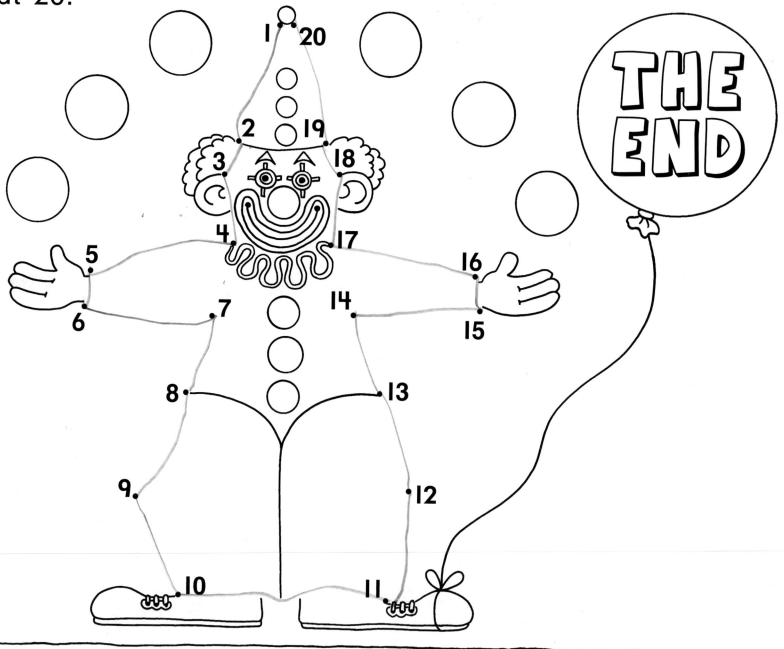

THE END

Say, **Draw lines to join the dots. Start at one and finish at twenty. Count carefully.**
Extension: Ask the children to read the message on the clown's balloon.